Liaison Functions

Held at the 2023 AIChE Annual Meeting

Orlando, Florida, USA
5-10 November 2023

ISBN: 978-1-7138-9320-2

Printed from e-media with permission by:

Curran Associates, Inc.
57 Morehouse Lane
Red Hook, NY 12571

Some format issues inherent in the e-media version may also appear in this print version.

Printed with permission by Curran Associates, Inc. (2024)

AIChE
120 Wall Street, FL 23
New York, NY 10005-4020

Phone: (800) 242-4363
Fax: (203) 775-5177

www.aiche.org

Additional copies of this publication are available from:

Curran Associates, Inc.
57 Morehouse Lane
Red Hook, NY 12571 USA
Phone: 845-758-0400
Fax: 845-758-2633
Email: curran@proceedings.com
Web: www.proceedings.com

TABLE OF CONTENTS

AICHE JOURNAL FUTURES: NEW DIRECTIONS IN CHEMICAL ENGINEERING RESEARCH (INVITED TALKS)

ANDREAS ACRIVOS AWARD FOR PROFESSIONAL PROGRESS IN CHEMICAL ENGINEERING LECTURE

EMERGING JUNIOR INVESTIGATOR OPEN INNOVATION FORUM (INVITED TALKS)

JOHN M. PRAUSNITZ AICHE INSTITUTE LECTURE

KICHE-US CHAPTER OPEN FORUM (INVITED TALKS)

SBE'S JAMES E. BAILEY AWARD LECTURE

THE LANGER PRIZE FOR INNOVATION AND ENTREPRENEURIAL EXCELLENCE AWARD PRESENTATION AND LECTURE

WILLIAM R. SCHOWALTER LECTURE

SBE AWARDS LECTURES - D.I.C. WANG AND JUNE WISPELWEY BIO LEADERSHIP FELLOWSHIP

HIGHLY EFFICIENT DOE FOR RAPID PROCESS DEVELOPMENT

HOW TO MAINTAIN PRODUCTIVITY IN A CRISIS

NETWORKING FOR NERDS: FIND AND FORGE RELATIONSHIPS THAT DELIVER ROI FOR ALL

67a Engineering *Vibrio natriegens* for Degrading and Assimilating Poly(ethylene terephthalate)

Tianyu Li, Stefano Menegatti, Nathan Crook

Poly(ethylene terephthalate) (PET) is a highly recyclable plastic material that has been extensively used and manufactured. Like other plastics, PET resists natural degradation, therefore accumulates in the natural environment. Several recycling strategies have been applied to manage waste PET, but these tend to result in downcycled products that eventually end up in landfills. This accumulation of landfilled PET waste contributes to the formation of microplastics by being broken down into small pieces that pose a serious threat to marine life and ecosystems, and potentially to human health. To address this issue, our project leverages synthetic biology and metabolic engineering to develop a whole-cell biocatalyst capable of degrading and assimilating waste PET in seawater environments. Specifically, we focus on using the fast-growing, nonpathogenic, moderate halophile *Vibrio natriegens* to construct this biocatalyst. Our talk will highlight two main processes we implemented in our bacterium: PET depolymerization and metabolism of PET breakdown products.

67b Applications of Lignin-derived Deep Eutectic Solvents in Biorefinery Processes

Chang Geun Yoo, Yunxuan Wang, Jiae Ryu, Kwang Ho Kim, Xianzhi Meng, Yunqiao Pu, Yang Tian, Aymerick Eudes, Gyu Leem, Arthur Ragauskas

Chang Geun Yoo,[1,9] Yunxuan Wang,[1,2] Jiae Ryu,[1] Kwang Ho Kim,[3] Xianzhi Meng,[2] Yunqiao Pu,[4,5] Yang Tian,[6,7] Aymerick Eudes,[6,7] Gyu Leem,[8,9] Arthur J. Ragauskas[2,5,10]

[1] Department of Chemical Engineering, State University of New York Environmental Science and Forestry, Syracuse, NY 13210, United States

[2] Department of Chemical & Biomolecular Engineering, University of Tennessee-Knoxville, Knoxville, TN 37996, United States

[3] Clean Energy Research Center, Korea Institute of Science and Technology, Seoul, 02792 Republic of Korea

[4] Biosciences Division, Oak Ridge National Laboratory, Oak Ridge, TN 37831, United States

[5] Joint Institute for Biological Sciences, Oak Ridge National Laboratory, Oak Ridge, TN 37831, United States

[6] Environmental Genomics and Systems Biology Division, Lawrence Berkeley National Laboratory, Berkeley, CA 94720, United States

[7] Joint BioEnergy Institute, Emeryville, CA 94608, United States

[8] Department of Chemistry, State University of New York Environmental Science and Forestry, Syracuse, NY 13210, United States

[9] The Michael M. Szwarc Polymer Research Institute, Syracuse, NY 13210, United States

[10] Center of Renewable Carbon, Department of Forestry, Wildlife, and Fisheries, University of Tennessee Institute of Agriculture, Knoxville, TN 37996, United States

The production and use of organic solvents from fossil fuels pose challenges in the sustainability field, including the effort and energy required for their disposal. There have been enormous efforts to seek alternatives to these types of media in many research areas. Deep eutectic solvents (DESs) have been recently introduced as a promising solvent due to their great potential including low vapor pressure, high boiling point, biodegradability, and high reactivity.

Renewable components have been studied as hydrogen bond donors in many DESs, and these solvents are used for biofuel production, metal electrodeposition, electrolyte, separation and other application areas. This talk will introduce sustainable biorefinery processes using lignin-derived DESs and their possible applications in the production of biofuels and chemicals. A closed-loop strategy using these renewable DESs to achieve a circular bioeconomy will be discussed.

67c Predicting Thermodynamic and Transport Properties of Organophosphates: A Molecular Simulation Substitute for Experiments that You Don't Want to Perform

Gennady Gor, Ella Ivanova

While the production and stockpiling of organophosphorus chemical warfare agents (CWAs), such as sarin, was banned three decades ago, CWAs have remained a threat. New approaches for decontamination and destruction of CWAs require detailed knowledge of their various physicochemical properties. Organophosphorus CWAs, are colloquially called 'nerve gases', however they are stockpiled as liquids and form aerosols when dispersed. To predict the formation and evolution of CWA aerosols one needs to know such properties as vapor pressures, surface tension and viscosity.

Due to the extreme toxicity of sarin, most experimental studies are carried out using its surrogates -- organophosphorus compounds which, while having similar structures, are much less toxic, e.g., dimethyl methylphosphonate (DMMP) and diisopropyl methylphosphonate (DIMP). However, not only for sarin, but also for its surrogates, literature data on the surface tension and viscosity are scarce. Therefore, theoretical predictions based on molecular dynamics simulations can complement and even substitute experimental data.

Here we present predictions of the surface tension and viscosity of sarin and its two common surrogates, DMMP and DIMP from the classical molecular dynamics simulations. We utilized the Transferable Potentials for Phase Equilibria (TraPPE) force field, which demonstrated an excellent agreement with the available experimental data for both properties. Our simulations provided the surface tension and viscosity data in the broader range of temperatures than what was measured experimentally. Furthermore, these data allowed us to evaluate suitability of different surrogates. Our results suggest that for applications where liquid viscosity is important, DMMP is a more suitable surrogate for sarin than DIMP. However, it is not the case when surface tension is more important, despite the difference in molecular size, that surface tension of sarin is closer to that of DIMP than to DMMP. The temperature-dependent surface tension and viscosity values for sarin and its surrogates obtained in our study can be used in large scale models predicting the formation and evolution of their aerosols.

67d Leveraging multiplex genome engineering and metabolic pathway design to upcycle waste plastics and biomass via aldehyde transformations in live bacteria

Aditya Kunjapur

The interconversion of chemical functional groups on building block molecules can broaden their properties and therefore their value as synthons and monomers for macromolecules. Carboxylic acids are an example class of molecules commonly found in the chemical or biological deconstruction products of waste plastic and biomass, and they serve as excellent targets for selective functional group interconversions. One of the most versatile classes of functional groups for the formation of C-C or C-N bonds are aldehydes, which can be biosynthesized from carboxylic acids via carboxylic acid reductase (CAR) enzymes. We recently exploited the breadth of CAR substrate specificity to design a route from terephthalic acid (TPA), a deconstruction product of polyethylene terephthalate plastic (PET), to a diamine via terephthalaldehyde (TPAL) using purified enzymes [1]. In our submission to the AIChE Futures Issue, we investigated whether we could conduct this biochemical transformation in an aerobic fermentation using engineered *Escherichia coli* [2]. Despite our prior efforts to limit the reduction of aldehydes to their corresponding alcohols in *E. coli* by performing gene deletions [3], to our surprise we found that TPAL was one of the only aldehydes that we have tested that remained subject to rapid reduction when added to cultures of our previously engineered strain. In this talk, I will present our efforts to chase down and eliminate the remaining aldehyde reductases responsible for the reduction of TPAL, which included one gene that had not previously been targeted in any aldehyde stabilization effort to our knowledge. I will then describe the various transformations that we could perform when supplementing TPAL to cultures of these cells in aerobic fermentation. These include amine products as well as a non-standard amino acid that we recently reported in another publication [4]. If time permits, I will also mention our surprising discovery that aldehyde oxidation, rather than reduction, is catalyzed by resting whole cell biocatalysts, which we have also addressed using genome engineering [5]. Finally, we can employ analogous strategies to obtain amines from biomass-derived carboxylic acids in addition to plastic-derived carboxylic acids [6]. Collectively, these strategies expand our design space for transformations that are challenging using traditional synthetic approaches and for which we are using low-cost biocatalysts and sustainable feedstocks.

[1] Gopal et al. ACS Catal. 2023 13 (7), 4778-4789.

[2] Dickey et al. AIChE J. 2023.

[3] Kunjapur et al. JACS. 2014 136 (33), 11644-11654.

[4] Jones et al. Commun. Biol. 2023.

[5] Butler et al. Metab. Eng. 2023 77, 294-305.

[6] Nain et al. BioRxiv. 2023.

67e Surface Anisotropic Particles, Capillary Interactions, and Dynamics of Fluid Interfaces

Elton Lima Correia, Sepideh Razavi

Presence of complex solutions composed of fluids, ions, surfactant molecules, and colloidal particles is commonplace in problems relevant to materials discovery and manufacturing. Such multicomponent fluidic systems are often confined by interfaces in processes associated with the water-energy nexus as well, for example, in membrane separations and subsurface energy recovery and storage. To make matters more intricate, fluid interfaces are not static and are constantly subject to external disturbances such as thermal gradients, imposed stresses, and changes in composition. Given the environmental and economic impact of the subject matter, it is important to advance our fundamental quantitative understanding of the complex interfacial systems just summarized, with the goal of ultimately predicting and controlling their behavior in relevant high-tech applications. In this presentation, I will review our recent findings on how particle surface anisotropy influences the stability and rheology of fluid interfaces. Our investigations demonstrate that particles' amphiphilicity is a crucial factor governing their orientation at the interface, which in turn dictates the complexity of capillary interactions present at a fluid interface and its ensuing mechanical properties. Understanding these concepts is key in designing interfacial systems, by engineering the surface attributes of particles, and is essential for numerous applications of soft matter, as listed in the New Directions for Chemical Engineering.

67f Designing better drug delivery particles: Protein corona formation and multicomponent aggregation of particles in blood plasma

Aida Lopez Ruiz, Guangliang Liu, Kathleen McEnnis

Drug delivery particles must overcome many biological barriers to successfully reach their target. One of those challenges is to evade the immune system and circulate through the body until they find their intended target. Many different engineered particles have been used to improve the circulation time of particles such as by adding poly(ethylene glycol) (PEG) ligands, adding zwitterionic polymers, adjusting particle size and shape, and attaching peptides to the surface. Though some of these tactics have resulted in longer circulation times, there is still significant off-site accumulation of the particles. It is difficult to determine why off-site accumulation occurs and how changes to the particle surface affect it as there are very few tools available to directly study particles in blood.

Characterizing the size of the particles directly in blood plasma is important to determine the extent of protein corona formation on the particles. With DLS, this is typically done by exposing particles to blood plasma followed by washing or dilution with water, then the increase in particle size can be measured. By washing the particles, however, the loosely bound proteins of the soft corona are removed and the measurements do not reflect what the particle would experience *in vivo*. As an alternative, we have developed a method of analyzing particles in blood plasma using nanoparticle tracking analysis (NTA) using fluorescently labeled particles. The size of the particle, and thus the size of the protein corona, can be measured in pure blood plasma using this method. Additionally, due to the unique visualization of the nanoparticles, the aggregation behavior of the particles in blood plasma can be observed and quantified. In this work, polystyrene particles were modified with multi-arm and linear polyethylene glycol (PEG) to determine their aggregation profiles and protein corona formation. Multi-arm PEGs were found to aggregate more than linear PEGs, due to the change in zeta potential from unreacted end groups, which may lead to shorter circulation half-lives. Furthermore, the protein corona formation and composition were studied after different washing procedures, highlighting the importance of studying protein corona formation with undiluted blood plasma.

67h Stability and kinetics of Fe-based MOFs with diverse structures in aqueous pollutant degradation

Samuel C. Moore, Michele Sarazen

Global access to clean water is an essential and growing need due to concomitant increased water consumption and pollution. Thus, advanced, effective materials, such as metal-organic frameworks (MOFs), that can remediate water contaminates either through their absorption or degradation are of mounting importance. MOFs made from earth-abundant metals, particularly iron (Fe), are attractive given their low toxicity and metal cost and have demonstrated water pollutant degradation at moderate conditions in catalytic and photocatalytic processes. This work focuses on hydrogen peroxide oxidation of methylene blue, a probe molecule used to understand catalyst performance in the context of the breakdown of medication and dyes from pharmaceutical and textile industries, respectively. We aim to elucidate structure-functional relationships driving trends in catalytic reactivity and stability on two polymorphic Fe-MOFs—MIL-101(Fe) and MOF-235. Under excess oxidant conditions, MIL-101(Fe) demonstrates a larger lumped first order rate constant (normalized by total Fe) than MOF-235 with the same secondary building units in different crystallographic arrangement, indicating the role disparate ligands coordinated to metal nodes play in reactivity. However, significant structural changes are evident, including loss in crystallinity and apparent Fe leaching, even though the catalysts continue to demonstrate degradative reactivity upon recovery and reuse. Moreover, leaching occurs even when catalyst is exposed to water in the absence of reactants; thus, solvent interactions drive material changes that require careful interpretation to clarify the nature of active sites and reactivity. MOF stability is extended to other systems, probing effects of different solvents (polar, aprotic (acetonitrile) and polar, protic (water)) on deactivation and dominant surface species. This work exemplifies the need to understand the timescale and extent of structural changes during reaction to properly contextualize material performance across aqueous oxidation catalysts.

214a Charge Neutral Polymer Complexes as Battery Components

Rachel Segalman

Polymeric components impart inherent mechanical durability and decreased flammability to electrochemical devices, but must have higher performance metrics in order to gain widespread use. This is a particular challenge as long-range Lithium ion transport is directly related to the viscosity matrix material and is ultimately limited by the sluggish dynamics of polymers. In this talk, I will discuss superionic conductivity in which the ion motion is decoupled from matrix dynamics and instead occurs through free volume elements in the structure. I will discuss our work to design semi-crystalline zwitterionic polymers that demonstrate superionic conductivity and also very high salt solubilities due to their polarizability. As a result, they have both high Lithium ion conductivities (10-3 S/cm) and cation transport numbers (t+=0.67) despite their modest glass transition temperatures (0-25°C). I will also discuss new coacervate-inspired battery binders that demonstrate both high ion and electron conductivities and their use to in composite electrodes compatible with polymer electrolytes.

81a Beyond Surface Facilitates Electrocatalytic Reactions of Renewable Carbons

Dohyung Kim

There has been growing interest to drive energy and chemical transformations via the use of renewable electricity to address sustainability challenges. The success of the approach rests on the use of the right materials to efficiently catalyze chemical reactions and there have been intense efforts to develop advanced electrocatalysts. Despite the progress, there is still much room for improvement. It is because of how we typically view catalytic reactions at the solid-liquid interface that often lacks consideration of the liquid phase (e.g., solvent molecules, double-layer ions). The BEYOND SURFACE approach not only recognizes the presence and role of liquid phase components but alters their characteristics to facilitate chemical reactions. In this talk, examples of the BEYOND SURFACE approach are presented for electrocatalytic reactions of renewable carbons, CO_2 and biomass. The first example concerns the discovery of a unique interfacial configuration composed of a metal surface and surfactants that can promote reactions by tuning the electric double layer. The other example presents the need for a better understanding of solvent molecule behavior at electrochemically active interfaces and how such knowledge can be leveraged to achieve better activity for a catalytic reaction. These studies highlight the complexity of electrochemical interfaces and the potential of thinking beyond the traditional perspectives.

81b Rationally Designing Nanoporous Catalytic Materials for Sustainable Energy Applications

Hong Je Cho

The increasing demand for energy and commodity chemicals combined with the diminishing fossil fuel supply has resulted in an accelerated research focus on discovering new catalytic processes. Utilizing lignocellulosic biomass holds great promise for producing sustainable chemicals and fuels. In addition, tandem catalytic processes, which allow multi-step reactions to occur in a single vessel, offer an alternative strategy to address the challenge by eliminating costly separation steps and increasing energy efficiency. Catalysts have been a key enabler in identifying and navigating sustainable and efficient energy processing.

In this talk, I will present how nanoporous catalytic materials (in particular, zeolites) should be rationally designed for biomass conversion and tandem catalysis. The scientific implications of this research are significant for the sustainable and efficient production of chemicals and fuels. Such knowledge has built a solid foundation for tailoring nanostructured catalytic materials to tackle global energy challenges.

81c Nanoparticle-Based Crosslinks in Polymer Networks

S. Eileen Seo

Inherent self-healing properties make materials formed from noncovalent networks attractive as they can repair fractured surfaces facilitated by highly mobile chain segments and by possibilities of achieving fast reversible associations at ambient conditions. However, noncovalent networks are prone to creep and considerable material deformation on loading, which makes these materials unsuitable for many applications. Stronger bond strengths lead to better mechanical properties, but the reaction kinetics of reversible bonds become slower and self-healing becomes less efficient. Multivalent binding of ligands on one entity via noncovalent interactions is collectively much stronger and display qualitatively different properties than those of underlying monovalent interactions. Here, we introduce the concept of multivalency in crosslinking mechanisms using surface-functionalized nanoparticles to make noncovalent networks with improved mechanical properties. First, a new type of dynamic, noncovalent networks with enhanced bond strength is developed by introducing the concept of particle-based crosslinks and systematically studying the effect of positive cooperativity in polymer networks. Secondly, these materials are designed to have access to *two or more thermodynamic states* where the bond dynamics of polymer networks are reversibly programmed using tunable interfacial interactions between nanoparticles and polymers. The underlying hypothesis is that the number of binding events between nanoparticles and polymers and the strength of interfacial interactions exhibit non-linear relationship, which will be tested by measuring the crosslink's bond exchange rate and bulk mechanical properties.

81d Metabolomics-Guided Discovery of Metabolic Design Principles and Engineering Strategies

Junyoung Park

Metabolism is a complex network of biochemical reactions. Metabolism fuels all living organisms and plays a major role in the global carbon cycle. The past decade has witnessed the emergence of metabolomics as an essential tool for determination of the metabolic states of widely divergent organisms. The advancement and wide adoption of analytical tools such as mass spectrometry and nuclear magnetic resonance spectroscopy enabled comprehensive and sensitive quantitation of metabolites. However, a gap remains between metabolite measurement and rational engineering of metabolism. My lab aims to bridge the gap. Using state-of-the-art mass spectrometry, isotope tracing, and quantitative flux modeling, we uncover metabolic design principles and apply fundamental understanding to address challenges in energy and medicine. In this talk, I will highlight our recent investigation into central carbon metabolism and demonstrate how it can be tamed for successful application in biotechnology.

77a Translation of Biomedical Microtechnologies from the Lab to the Clinic

Mark Prausnitz

Many tissue barriers that limit access to the body exist on the micron length scale. We are therefore developing microtechnologies that selectively cross these barriers in a minimally invasive way to improve drug delivery and other medical applications. In some cases, this approach can target drug delivery to specific tissues. In other cases, this approach simplifies drug delivery for increased patient access and acceptability. In all cases, we emphasize technologies that are simple to use, are inexpensive to produce, and improve safety, efficacy and/or access to medical treatments.

Vaccination generally requires injection by needle and syringe. This method often limits access to vaccines in developing countries for lack of trained health care providers; limits compliance with influenza, COVID and other vaccinations due to patient hesitancy to be injected; limits safety because of inherent dangers of needlesticks and biohazardous sharps waste; and limits immunogenicity by vaccinating typically in the muscle, which often provides inferior immune responses compared to vaccination in skin and mucosal tissues. To address these limitations, we are developing microneedle patches for vaccination. These patches contain arrays of solid, conical microneedles made of a mixture of vaccine and safe, water-soluble materials. When pressed to the skin, the microneedles dissolve in the dermal fluids within a few minutes to release the vaccine. Microneedle patches can increase access to vaccination because little to no expertise is needed to administer them. They can improve vaccination compliance because they are painless and convenient. They can improve safety because the microneedles dissolve in the skin, leaving no sharps waste. And they can improve immune responses by targeting vaccine delivery to the immune cells of the skin. We have developed microneedle patches to achieve these various capabilities, and have developed them through design, fabrication, in vitro characterization, and in vivo immunogenicity studies, including human clinical trials.

Some vaccines, like the mRNA vaccines used to combat the COVID-19 pandemic, need to be delivered into cells. Current mRNA vaccines accomplish this using lipid nanoparticle delivery systems, but these formulations are complex, costly and may be associated with side effects. Electroporation is a well-established way to deliver nucleic acids into cells in the lab using short electric pulses, but its use for human vaccination is limited by high cost, complex instrumentation and painful delivery. We developed a novel electroporation device that generates pulses using a low-cost piezoelectric element combined with microneedle electrodes that limit the electric field to the cells found in the skin's epidermis. In this way, we have achieved electroporation that was effective to administer a DNA vaccine against SARS-CoV-2 in mice using a device expected to cost less than $1.00 that was reported as painless by human subjects.

Microneedle patches are effective to breach the skin barrier but are difficult to use on large areas of skin, as might be needed in dermatology or cosmetics. We therefore developed STAR particles, which are millimeter-scale particles with microneedles sticking out. When rubbed on the skin, the STAR particles make micropores in the skin through which drugs can diffuse. STAR particles were effective to increase skin permeability to dermatological drugs by one to two orders of magnitude in vitro, improved treatment of melanoma by topical chemotherapy in mice, and were found to be painless and well-tolerated in human subjects, even after repeated use.

Most biomarkers of use in clinical medicine are collected from blood, but blood is only about 7% of bodily fluids. Interstitial fluid (ISF), which is the extracellular fluid bathing cells in tissues, is more than 20% of bodily fluid and contains both systemic as well as local tissue biomarkers. It also does not clot or contain cells, which facilitates continuous or repeated measurement and simplifies sample preparation before analysis, respectively. However, ISF is difficult to obtain because its collection is limited by flow through a nanoporous extracellular matrix, and its extraction involves tissue access that generally causes bleeding. Building off our use of microneedles to create transport pathways for drug delivery into the body, we have adapted microneedles to work "in reverse" and pull material out. The combination of microneedle skin puncture followed by application of suction has enabled us to collect microliter quantities of ISF without bleeding in human subjects. ISF was found to contain almost all of the biomarkers found in companion blood samples, as well as hundreds of additional compounds not found in the blood. This technology may enable ISF to become a novel source of biomarkers for research and clinical medicine.

Microneedle technologies are not only useful in the skin, but also for targeted drug delivery in the eye. Many drugs benefit from delivery directly to the site of their action so as to avoid side effects at off-target locations. We developed a hollow microneedle – much like a tiny hypodermic needle – that penetrates less than 1 mm into the ocular surface to reach a part of the eye called the suprachoroidal space. Injection into this space sequesters the drug in the choroid and adjacent retina, which is the site of action for most drugs given to the back of the eye. This delivery technology has been developed into a product that administers triamcinolone acetonide, which is a steroid that treats inflammation in the back of the eye. This product was approved by FDA in late 2021 and has been made available to patients since early 2022.

170a Intelligent E-Nose System: Materials Discovery to Wearable Device

Nosang Myung

Electronic detection of gaseous molecules is rapidly emerging as an alternative to the conventional detection methods because of the small size, low-power consumption, and improved sensing performance. In this presentation, we will discuss the development of smart electronic-nose (E-nose) systems with superior sensing performance. To mimic biological olfactory systems, we optimized the high-density sensor array architecture by vertical integration of unique combinations of libraries of nanoengineered materials to identify and quantify target analytes where the sensing performance was maximized by nano- and defect- engineering of materials. Once sensor response data has been collected using high-throughput screening systems, machine learning techniques were utilized to select the optimal subset of sensor elements for detecting analytes. Manufacturability of devices were demonstrated by prototyping a wearable E-nose watch using a combination of industry-standard printed circuit board manufacturing techniques and proven, scalable, additive manufacturing techniques developed in our lab.

170b Visualizing Large-Scale Biological Conversion Processes in Real-Time

Taewook Kang

The volumetric mass transfer coefficient of gas molecules is one of critical factors when designing and scaling up gas-liquid reactors as well as improving the efficiency of gas conversion. This is typically done by chromatographic techniques. For example, sampling a gas molecule of interest from the headspace of the gas-liquid reactor and subsequent gas chromatography measurement with Henry's constants allows us to estimate the amount of the gas molecule such as carbon monoxide dissolving in water. However, this sampling-based method for determining the volumetric mass transfer coefficient is only limited to the gas-liquid reactors with relatively large gas and liquid volumes. In this talk, I will report how to precisely determine volumetric mass transfer coefficient (kLa) of carbon monoxide (CO) in miniaturized gas-liquid reactors via sensitive probing of the Raman transitions of aqueous trace CO without sampling. Then, I will briefly discuss our recent efforts to visualize biological conversion processes that take place in large-scale bioreactors in real-time. I am confident that our proposed method has a huge impact on biological conversions ranging from design and scale-up of gas-liquid reactors, process monitoring and optimization to high-throughput screening of strains or catalysts.

170c Role of Process Systems Engineering (PSE) in Strategizing the Path to Carbon Neutrality

Jay Hyung Lee

Despite the clear mandate towards the goal of carbon neutrality, the current pace of progress for the transition to renewable energy and the decarbonization of chemicals production is rather slow. The key challenge facing government and industry is identifying low-carbon solutions that are economically feasible yet impactful in reducing greenhouse gas (GHG) emissions. Process systems engineering (PSE) is a field of research that applies methodological guidelines and systematic tools for process synthesis, design, operation, and evaluation. In this regard, PSE plays a critical role as an intermediary, bridging bench-scale R&D to higher-level policy and deployment decisions. In this talk, we examine the various paths to achieving carbon-neutrality including recycling, carbon capture and conversion, (CCU) and bioconversion, and important factors for them to succeed. We discuss the role of PSE tools, e.g., modeling, optimization, and life-cycle analysis, to assess their feasibilities and to build the most realistic path forward by combining them. Computer-aided tools for analysis and synthesis can be extremely helpful and a recently developed software tool for this purpose named ArKaTAC3 will be introduced.

170d Novel but Simple Method to Prepare Vanadia Catalyst with Superior Sulphur Resistance and NOx Removal Activity

Do Heui Kim

Nitrogen oxides (NOx) are a pollutant present in combustion gases that have serious deleterious effects on the human body, and these species also participate in photochemical processes that result in smog and acid rain. Nevertheless, NOx can be efficiently removed from exhaust gases by implementing selective catalytic reduction with NH_3 (NH_3-SCR). TiO_2-supported vanadia catalysts are the most common materials used in NH_3-SCR processes since they offer excellent denitrification ability and high sulfur resistance, in the temperature range from 300 to 400 °C. However, as environmental regulations become more stringent and applied to various fields in the future, the necessity of operating SCR at low-temperature below 250 °C becomes more important. Although it is known that V-based catalysts have relatively high sulfur resistance, they are not free from sulfur deactivation at the temperature below 250 °C due to the formation of ammonium bisulfate (ABS), which can physically block the pores in the catalyst, deteriorate the catalytic performance (Eq 1).

$NH_3 + H_2O + SO_3 = NH_4HSO_4$ (l) ---- (1)

Thus, the current technology has a huge dilemma in achieving both low-temperature activity and high sulfur resistance. Here, we report that Y zeolite can be physically mixed with the vanadia catalyst to effectively trap liquid ABS, protecting most V sites from deactivation by ABS and demonstrating stable NH_3-SCR performance at 220 °C. The sulfur resistance of physically mixed catalyst was 3 times higher than VWTi, and order of magnitude higher compared to the commercial low-temperature catalysts (Cu-SSZ-13 and Mn/TiO_2). Moreover, the activity of VWTi+ Z recovered almost completely after regeneration at 350 °C where ABS decomposes

This study proposed a simple but novel "physical mixing" strategy to achieve excellent sulfur resistance of the vanadia NH_3-SCR catalyst with high NOx removal ability. Moreover, the developed physical mixing catalyst was transferred to the POSCO steelmaking company and successfully commercialized in their sintering process.

170e Effect of Interfacial Energy and Geometry to Permeation of Liquid through Filter Pores

Dongjin Seo

The permeation of liquids through filter pores is studied as a function of contact angle of liquid and pore geometry. While it is generally known that the increased wettability, thus lower contact angles, increases the permeability, the quantification of its effect on permeability lacks. In addition, the effect of the shape/geometry of the pores on the permeation is not investigated. Yet, the geometry, especially the angle of the pore wall would also control permeability because it would change the radius of curvature of the menisci that would form inside the pores. This causes the difference in Laplace pressure and the different areas for a liquid to wet. Therefore, the dependence on permeability as a function of contact angle and pore geometry is studied.

To do so, an equation was developed that enables the calculation of the pressure for a liquid species, i, to permeate through a pore with the wall angle at α_j and the contact angle of the species i on the filter material, θ_i.

To verify the effect from the contact angle of a species through pores, θ_i, a cylindrical hole was prepared using a 3D printer, after which it was coated with PDMS to decrease the roughness. The contact angle of water on PDMS was altered by exposing the pore to UV-ozone. A glass tube was placed upon the pore and water was added gradually to observe the height at which the liquid started to permeate. The results that related the height to permeate as a function of θ_i was found to have the same functional form as the equation developed. To investigate the effect of α_j, tapered holes with various α_j were printed and used without further altering the contact angle of water ($\theta_{water} = 61°$). It is observed that the decrease in the angle increased the height to permeate.

170f Crystalgpt: Enhancing System-to-System Model Interchangeability in Crystallization Prediction and Control Using a Time-Series-Transformer Model

Joseph Kwon

Given the complexity of first-principled crystallization models, often a surrogate model is constructed to provide a computationally leaner digital twin for process monitoring and control. Although various state-of-the-art (SOTA) machine learning (ML) have been used as surrogates with high prediction accuracy, they often are very system-specific, and cannot easily be repurposed from system `A' to system `B'. For example, a plant with N different crystal systems will develop and maintain N bespoke surrogate models that show poor interchangeability despite the presence of significant commonalities in the general crystallization process. To address this issue, we took inspiration from the disruptive emergence of large-language models (LLMs) that leverage powerful transfer learning capabilities of transformers to generalize across different languages or tasks. To this end, a novel time-series-transformer (TST) framework with varying numbers of encoders, decoders, and attention heads was constructed.

Further, observable process data from N (i.e., 10 to 20+) different crystallizers, each with 5000+ operation conditions, was collected. Next, this large corpus of data was utilized to develop a single TST model (i.e., CrystalGPT) to act as a unified surrogate model for any of N systems, and even for a new $N+1^{th}$ crystal system. Further, CrystalGPT was integrated with a model predictive controller (MPC) to perform setpoint tracking for a batch crystallizer for a chosen crystal system K. Powerful predictive capabilities of CrystalGPT are demonstrated with a combined normalized-mean-squared-error (NMSE) of 5×10^{-4} over 10M+ data points, thereby showcasing an 8 to 10 times better performance than current SOTA ML models. Overall, the current work focuses on the development of a plant-wide surrogate model that leverages TST's transfer learning capabilities to seamlessly generalize across [N^{th}, $N+1^{th}$, $N+2^{th}$, ...] crystal systems with high-value implications in process monitoring and control.

382a Can biotechnology deliver cost effective fuels with reduced carbon footprint?

Gregory Stephanopoulos

The importance of liquid fuels in transportation is well established, yet, there are presently no viable options for their cost-effective production from renewable feedstocks. During the past 15 years we have been developing in my lab a system for the conversion of gas mixtures of hydrogen (or CO) and CO_2 to oils. The system comprises anaerobic fixation of CO_2 and conversion of the CO_2 fixation product (for example, acetate) to lipids, from which biodiesel, green diesel or Sustainable Aviation Fuels (SAF) can be produced. In another application, the CO_2 fixation product is converted to alkanes. Our work includes both the engineering of the microbes *and* development of a process to achieve gas to liquid conversion in prototype systems. These systems are scalable, make no use of land (beyond what is needed for generating renewable electricity for hydrogen production), do not compete with food and are cost competitive based on high level and TEA based cost analysis. I will present the essential features of this process in my talk; full details can be found in 5 papers cited below.

Just 5 selected references

1. **Peng Hu, *et al.*,** "Integrated system for biological conversion of gaseous substrates to lipids," *Proceedings of the National Academy of Sciences,* 201516867; DOI: 10.1073/pnas.1516867113 (2016).
2. **Jingyang Xu, *et al.*,** "Application of metabolic controls for maximization of lipid production in oleaginous yeast," *of the National Academy of Sciences,* 114(27): E5308-E5316; DOI: 10.1073/pnas.1703321114 (2017).
3. **J. Qiao, *et al.*,** "Rewiring metabolism to maximize lipid production in *Yarrowia lipolytica,"* *Nature Biotechnology,* 35: 173-177; DOI: 10.1038/nbt.3763 (2017).
4. **O. Park, *et al*,** "Synergistic substrate co-feeding stimulates reductive metabolism," *Nature Metabolism,* 1(6): 643–651; DOI: 10.1038/s42255-019-0077-0 (2019).
5. **Li, Jingbo,** *et al.,* "Synthesis of High-Titer Alka(e)Nes in *Yarrowia Lipolytica* Is Enabled by a Discovered Mechanism." *Nature Communications,* 11(1): 1-13.DOI: 10.1038/s41467-020-19995-0 (2020).

Selected for Editors' Highlights: www.nature.com/collections/idhhgedgigand.

2023 Langer Prize Winner Presentation: Pseudo-Random Collisions: From Data Storage to the Synthesis of Modified DNA

Albert J. Keung

There is a rapidly growing proportion of society positively impacted by the power to engineer DNA. Currently, the majority of applications are in native biological systems: in the development of therapeutics, in supporting basic and biomedical research and development, and in promoting the health of agricultural organisms. Yet, there are also intriguing abiotic impacts DNA could have including as the substrate for next generation data storage and computation. In this talk, I will describe some of the key questions we have tackled and new technologies developed that explore the power of DNA both within native biology and beyond. I will also focus on the role of random collisions of people, fields, and expertise in spurring much of the progress we have made. This includes our ongoing story of exploring the applied and commercial impact of technologies, and how we came to focus and distill a complex world of possibilities into the development of our first target commercial product.

171a Physically intelligent colloidal systems

Kathleen J. Stebe

Active colloids include bacteria which swim by action of flagella and biomimetic self-propelled colloids that move by consumption of chemical fuel. We have been developing the concept of *Active Surface Agents*, active colloids trapped at fluid interfaces to promote interfacial transport. Fluid interfaces are highly non-ideal, complex domains that impose constraints that alter swimming behavior. We study the bacterium *Pseudomonas Aeruginosa* (PA01) at interfaces and characterize several distinct swimming behaviors. We measure the flow generated by these swimmers using a recently developed flow visualization method *correlated displacement velocimetry*. The flow field has unexpected asymmetries whose structure we describe fundamentally using hydrodynamic theory. We explore the implications of our results on mixing in the interface and in the design of biomimetic systems. By understanding how biological swimmers move at fluid interfaces, we can develop design rules for artificial biomimetic systems to promote transport at fluid interfaces with broad implications in chemical engineering processes.

340a Synthetic Biology 2.0: the Dawn of a New Era

Huimin Zhao

Synthetic biology is the design of novel or improved biological systems using engineering principles, which has broad applications in medical, chemical, food, and agricultural industries. Thanks to the rapid advances in DNA sequencing and synthesis, genome editing, artificial intelligence/machine learning (AI/ML), and laboratory automation in the past two decades, synthetic biology has entered a new phase of exponential growth. In this talk, I will highlight our recent work in the development and application of novel foundational synthetic biology tools. Specifically, I will introduce two related stories. One is focused on development of an AI/ML enabled self-driving biofoundry for next-generation synthetic biology applications. The other is focused on design and creation of artificial photoenzymes capable of catalyzing abiological reactions that are extremely difficult or impossible to achieve by either naturally occurring enzymes or chemical catalysts.

96a Highly Efficient DOE for Rapid Process Development

Mark Anderson

To maximize your educational experience it will good that you follow along my presentation of case studies using DOE software. Prior to class, please download the software from www.statease.com/trial/. Also, download data files, the annotated presentation slides and the *Handbook for Experimenters* from www.statease.com/AIChE2023.

Via a series of case studies, this short-course teaches multifactor design-of-experiment (DOE) tools for process development. By strategically deploying DOE, chemical engineers can rapidly screen factors down to a vital few, characterize their main effects and interactions and achieve optimal settings for peak process performance. See how these tried-and-true statistical methods, made easy by software, empower experimenters to quickly converge on the "sweet" spot—factor settings that meet all product specifications at maximum yield and minimal cost. The presentation is geared to experimenters with only a basic knowledge of statistics, e.g., p-values, r-squared and regression modeling.

708a How to Maintain Productivity in a Crisis

Alaina Levine

Staying focused, clear, productive, and creative can be challenging at any time, given all of the many stimuli that we encounter every day. And of course, this is exacerbated when we are navigating a challenge or a crisis. In this session, we will discuss strategies and tactics to give you a boost and help you remain centered and mission-focused during any circumstance and especially in the middle of uncertainty. We will also address how to leverage the challenge to strengthen your resolve and skills and those of your team.

709a Networking for Nerds: Find and Forge Relationships that Deliver ROI for All

Alaina Levine

Networking is the most honorable and valuable endeavor in which you can engage, because it is built on a spirit of generosity. At its core, networking is all about crafting win-win alliances where both parties provide value. In this session, we will discuss strategies and tactics for finding new collaborators and building mutually-beneficial partnerships with professionals at in-person meetings and across the globe. We will practice our networking and you will discover actual wording and questions to use to spark meaningful conversations. We will also explore networking using digital platforms and how to engage with people you meet at virtual events. You will emerge more confident and comfortable in your networking- now that you know exactly how to do it. Let's network!

AUTHOR INDEX